What is Time?

Facts, Musings, and Speculations
About Time for Children

by

Peter I. Kattan

PETRA BOOKS
www.PetraBooks.com

What is Time?

Facts, Musings, and Speculations About Time for Children

www.PetraBooks.com

Email: info@PetraBooks.com

Printed in the United States of America

ISBN-13: 979-8-8691-6213-7

Introduction

Time is my favorite topic. I always wanted to write a book about time. Finally I decided to write one designed for children. This is not a book that teaches your child how to tell the time. There are numerous such books on the market. This book is different. Basically the book delves into the nature of time. The book is designed such that it includes facts, musings, and speculations about time. These are written in a very simplified form suitable for children. The topics covered are:

```
(1) Hours, Minutes, and Seconds,
(2) Years, Months, and Weeks,
(3) Past, Preset, and Future,
(4) The Calendar,
(5) Dimensions,
(6) The Fourth Dimension,
(7) Psychological Time,
(8) The Five Senses,
(9) Age of the Universe,
(10) Parallel Timelines,
(11) Time Travel,
(12) Mental Time Travel,
(13) The Future,
(14) Time and Dreams,
```

(15) Three Dimensions of Time,
(16) The Nature of Reality,
(17) Consciousness,
(18) Healing Yourself,
(19) Synchronicity,
(20) The End of Time,
(21) The Transformation,
(22) Time in the Bible.

The book includes 22 topics carefully written such that each topic appears on one single page. In addition, there are numerous images and quotations about time that are taken from various sources. It is recommended that children read first the 22 topics, and then proceed to the quotations. It can be noticed that the quotations are a lot more difficult reading for children.

The book may be difficult for children especially the second half of the book where we discuss some advanced concepts in the theory of time. This difficulty is even magnified when we consider that the latter parts of the book are more speculative than factual. But in these speculations we offer a hope that children (and most importantly adults) should be aware of.

1

Hours, Minutes, and Seconds

A day consists of 24 hours.

An hour consists of 60 minutes.

A minute consists of 60 seconds.

A day consists of 24 x 60 = 1440 minutes.

An hour consists of 60 x 60 = 3600 seconds.

A day consists of 24 x 60 x 60 = 86400 seconds.

24 hours is the time it takes the earth to make a complete rotation about its axis.

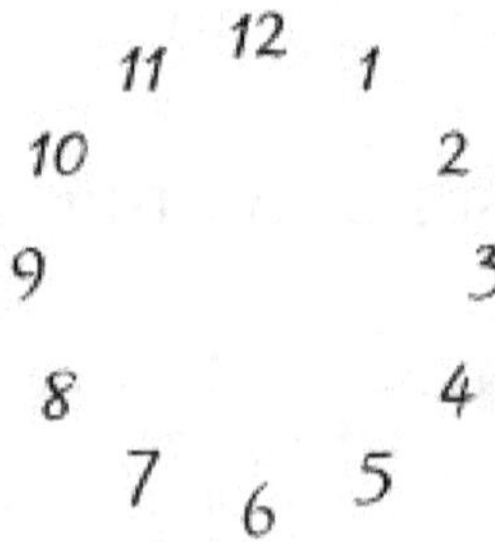

The idea that time flows … seems completely obvious, and is what you could call the common sense view of time. ….. But despite this, the idea that time is *not* linear, that in some way, the past, present, and future exist simultaneously, is not as nonsensical as it seems.

Steve Taylor, *Making Time: How Time Seems to Pass at Different Speeds and How to Control it* (pages 147-148)

2

Years, Months, and Weeks

A year consists of 52 weeks.

A week consists of 7 days.

The seven days of the week are Sunday, Monday, Tuesday, Wednesday, Thursday, Friday, and Saturday.

A year consists of 12 months.

Four months consist each of 30 days. These months are April, June, September, and November.

Seven months consist each of 31 days. These months are January, March, May, July, August, October, and December.

One month (February) consists of 28 days (29 days in leap years).

A year consists of 365 days (366 days in leap years). One year is the time it takes the earth to make a complete rotation around the sun.

In a computer game, there is no time. The game appears to have movement and progression due to two effects. One is the nature of our minds to process changing images into the concept of movement. The other is the progression of states of the game in terms of computer cycles…. This is exactly how time works according to conventional physics; that is, quantum mechanics.

Jim Elvidge, The Universe – Solved! (pages 200-201)

3

Past, Present, and Future

We live in the present.

What happened before now consists of the past. The past includes yesterday, last week, and years ago.

What will or may happen later consists of the future. The future includes tomorrow, next week, and years from now.

We know what happened in the past.

We know what is happening in the present (at our location).

We do not know what will happen in the future.

We study the past to learn from it and look forward to the future and make plans to better our life and others.

1 X 9 = 9	九	RAT	MID-NIGHT
2 X 9 = 18	人	OX	2AM
3 X 9 = 27	七	TIGER	4AM
4 X 9 = 36	六	HARE	6AM
5 X 9 = 45	五	DRAGON	8AM
6 X 9 = 54	四	SNAKE	10AM
1 X 9 = 9	九	HORSE	NOON
2 X 9 = 18	人	SHEEP	2PM
3 X 9 = 27	七	MONKEY	4PM
4 X 9 = 36	六	COCK	6PM
5 X 9 = 45	五	DOG	8PM
6 X 9 = 54	四	BOAR	10PM

Time is so fundamental in our experience of the world that any attempt to tinker with it meets with great skepticism and resistance.

Paul Davies, God and the New Physics (page 119)

4

The Calendar

A calendar is a system of organizing days for social, religious, commercial, or administrative purposes.

This is done by giving names to periods of time such as days, weeks, months, and years.

A calendar is also a physical device, often paper.

A lunar calendar follows the motion of the moon.

A solar calendar follows the motion of the sun.

The Gregorian calendar is the international standard that is used almost everywhere today.

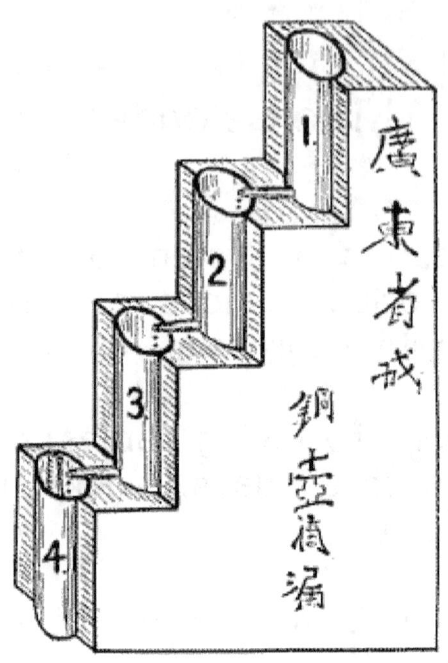

Time elapses more slowly for an individual in motion than it does for a stationary individual. … shouldn't one be able to live longer by being in motion rather than staying stationary?

Brian Greene, The Elegant Universe: Superstrings, Hidden Dimensions, and the Quest for the Ultimate Theory
(page 41).

5

Dimensions

We live in a world of three dimensions.

The first dimension is when you look in front of you and behind you.

The second dimension is when you look to your right or to your left.

The third dimension is when you look upward or downward.

We can locate any point in space by specifying its three dimensions.

Sometimes the three dimensions are represented by the letters x, y, and z.

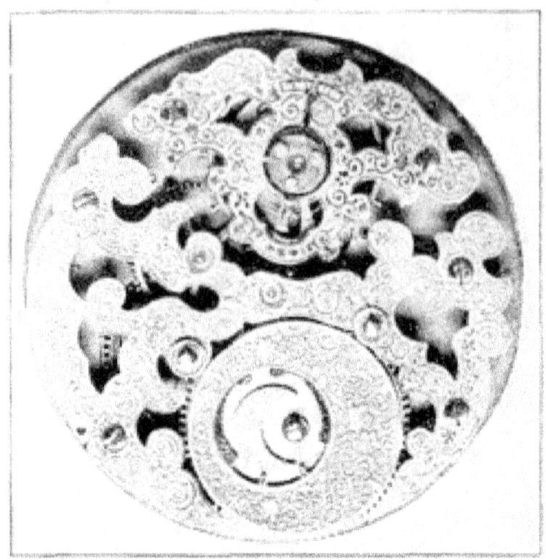

Time is not absolute but relative to the position of the observer. This idea, in the Special Theory, was expressed in two forms: first, in the time-dilatation effect; secondly, in its stress on the fact that two observers at different distances from an event will see it at different times.

J. B. Priestley, Man and Time (pages 88, 91)

6

The Fourth Dimension

Time is the fourth dimension according to Albert Einstein.

In his theory of relativity, Einstein treated time as a dimension like the other space dimensions, but slightly different.

Instead of treating space and time separately, Einstein considered them as one entity that he called the space-time continuum.

Like a location in space can be specified using three space dimensions, similarly an event in space-time can be specified using four dimensions: three of space and one of time.

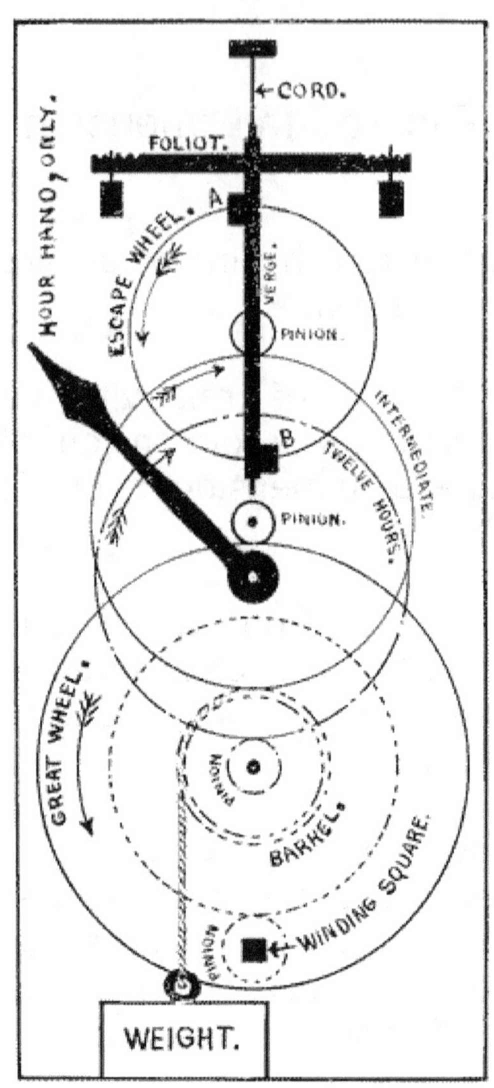
CORD.
FOLIOT.
HOUR HAND, ONLY.
ESCAPE WHEEL. A
VERGE.
PINION.
INTERMEDIATE.
B
TWELVE HOURS.
PINION.
GREAT WHEEL.
PINION
BARREL.
WINDING SQUARE.
PINION
C
WEIGHT.

7

Psychological Time

Sometimes we feel that time passes quickly but other times we feel that time passes slowly. This is what we mean by psychological time.

When we are having fun and enjoying ourselves, time flies and passes very quickly. In this case one hour will feel like ten minutes only.

However, when we are bored and doing something we do not enjoy, time creeps and slows down. In this case one hour will feel like ten hours.

In extreme cases, our perception of time changes completely. These include cases of accidents, near-death experiences, and meditation.

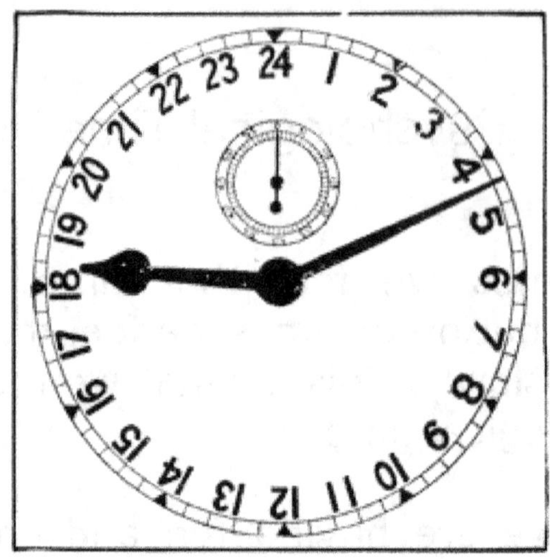

Saint Augustine once said: "What, then is time? I know well enough what it is, provided that nobody asks me; but if I am asked what it is and try to explain, I am baffled."

John R. Violette, Extra Dimensional Universe (page 50)

8

The Five Senses

We have five senses for seeing, hearing, touching, smelling, and tasting. Each of these senses has its own physical organ in the human body. For example, the eye is the organ used for seeing while the ear is the organ used for hearing.

However, we do not have a physical organ for sensing time. We feel the passage of time but do not know how.

Some scientists say that time is a construct of the mind. Without an existing mind (i.e. of an observer) there can be no time.

Thus, if there is a physical organ for sensing time, it will most probably be connected somehow with the brain – where the mind originates.

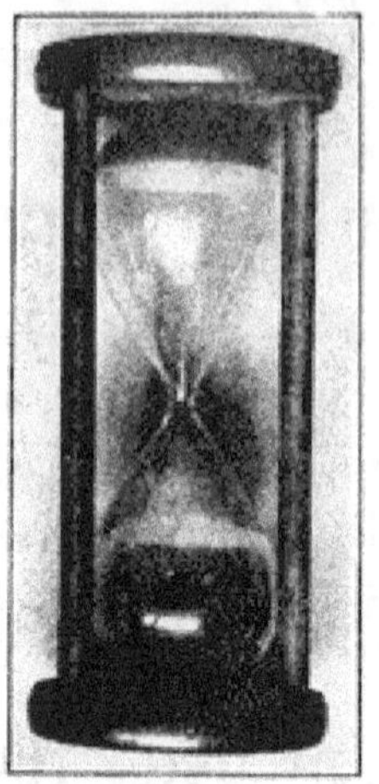

… why do the space and time dimension of our world begin to disintegrate at the particular level that they do? Why do dimensions no longer serve as sensory potentials at the magnitude of the Planck constant, rather than at some other level?

Samuel Avery, The Dimensional Structure of Consciousness (page 29)

9

Age of the Universe

Scientists say that the age of our universe is about 14 billion years. They say that the age of our earth is about 4 billion years. They determined this age based on what is called carbon dating.

During these millions of years, time passed. But how do we know that time passed millions of years ago? Clearly there was no mind at that time to sense time or count the hours and years.

So how can they say that time existed for millions of years without there being a mind (i.e. of an observer) for millions of years sensing time. Man with a mind first appeared only thousands of years ago. Clearly there was no man or mind millions of years ago.

There is no time without us.

Fig. 12   Fig. 13.

… the idea that time "flows," or that the present moment somehow moves from the past to the future in time, has no place in the physicist's description of the world.

Paul Davies and John Gribbin, *The Matter Myth: Dramatic Discoveries That Challenge Our Understanding of Physical Reality* (page 134)

10

Parallel Timelines

Scientists today believe in the existence of parallel universes, i.e. parallel timelines. These are universes occupying the same continuum as our universe but cannot be perceived by us using our normal five senses.

It is believed that a parallel timeline exists for each thought or decision we make. Suppose we decide to drink coffee instead of tea, then there exists a second parallel universe in which we drink tea instead of coffee.

Branching of parallel timelines often occurs when accidents or near accidents happen. For example, suppose you undergo a near accident today. Then, there are parallel timelines (universes) in which the accident occurs and you are injured, another one in which the accident occurs and you die. And so on.

There is currently no proof for the existence of these parallel timelines. But we seem to visit them when we experience some types of vivid dreams.

11

Time Travel

Scientists claim that time travel is possible or would be possible in the future. Traveling through time to the past or the future requires a time machine.

There are numerous books and papers detailing how to construct a time machine. However, this requires advanced technology and exotic materials.

Some scientists propose to use black holes to travel through time. However, this is beyond man's ability at the present time.

There are paradoxes associated with time travel. Suppose that you travel to the past and meet your grandfather. Then suppose that you kill your grandfather. If your grandfather was dead before you were born, how did you come into existence.

The above paradox can be resolved by using parallel timelines in which you time travel to a parallel universe different from the one in which you exist in now.

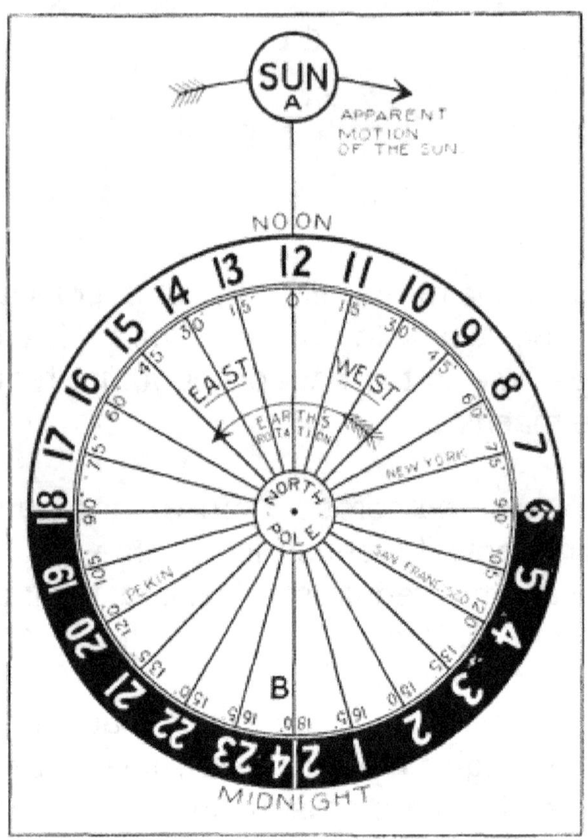

Libet … showed that conscious awareness only comes when a certain time period, around 500 ms, has passed. He calls this the time of "neuronal adequacy." If neuronal adequacy is not achieved, awareness does not take place. … It appears to the subject that no delay in awareness has occurred at all.

Fred Alan Wolf, *The Dreaming Universe: A Mind-Expanding Journey into the Realm Where Psyche and Physics Meet* (page 92-93)

12

Mental Time Travel

Another approach to time travel is to use the mind to travel into the past or the possible future. In this case the body remains in the present.

To use the mind to travel to the past, it is necessary to rely on vivid memories of that past. Intense visualization of past memories is considered some form of mental time travel.

To use the time to travel to the future, it is necessary to visualize a desired future event and concentrate on visualizing it intensely. You must focus solely on this possible future event.

When performing mental time travel, it is not enough to use images in the mind. These images may be accompanied by sound and other senses. Intense emotions are important too. In this regard listening to certain types of relevant music would be very helpful.

… we project fixity and absoluteness upon space and time where in fact there is none. … they are "essentially psychic in origin." Because of our absolutist habits we erect inner barriers to understanding synchronicity and lose our money to Spacetime Sam.

Victor Mansfield, Synchronicity, Science and Soul-Making (pages 92)

13

The Future

Can we predict the future? Is there free will? These are pertinent questions that need answers.

According to the latest theory of quantum mechanics, there are several probable futures and what happens depends on our present decisions and actions.

On the local level we do have free will and the future is not determined. However, on the global level, we do not have free will and the future is determined.

For example, consider a passenger on a plane flight across the Atlantic Ocean. On the local level (within the inside of the aeroplane) the passenger has free will in the sense that he can read whatever he or she wants and go or not go to the bathroom. However, on the global level, the passenger does not have free will in the sense that he cannot alter the path of the flight or force it to land somewhere else.

What is the connection between time and parallel universes? What is time? Having grasped time, what is space? … time turns out to be an imaginary dimension of space, or if you prefer, space is an imaginary dimension of time. In either case, the new view of time expands our awareness capability, making parallel universes more than just a thought.\

Fred Alan Wolf, Parallel Universes: The Search for Other Worlds (pages 112-113)

14

Time and Dreams

There are different kinds of dreams. Some dreams are wish fulfillment, other dreams are connected with something physical in the sleeper's environment, and some other dreams are telepathic.

However, a fourth kind of dream that interests us here are precognitive dreams. These are dreams in which you see an event from the future. These kinds of prophetic dreams are closely connected with time.

Precognitive dreams usually come true. However, in some cases the dream does not materialize. In these cases, the event in the dream does occur but in a parallel timeline.

Some types of vivid dreams are actually visits to parallel universes.

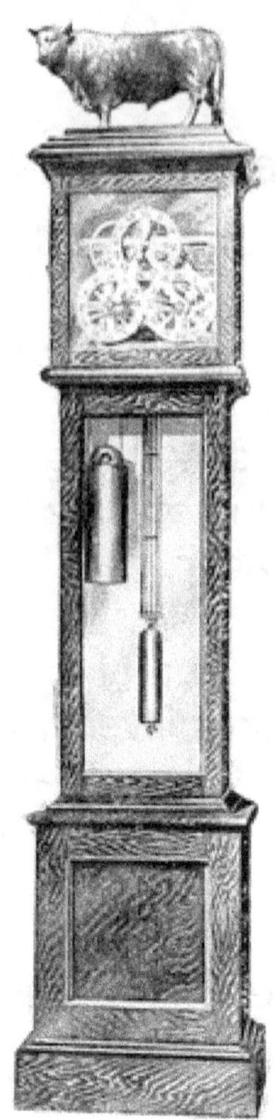

15

Three Dimensions of Time

Time seems to flow in one dimension. They talk about the river of time going from the past to the present to the future.

But some scientists propose the existence of three dimensions of time. The first dimension is the one we know in this timeline.

The second dimension of time is perpendicular to the first dimension – this means that the second time dimension flows across multiple timelines.

The third dimension of time is perpendicular to the other two dimensions. This dimension represents eternity.

We will not elaborate on the three dimensions of time except to say that their mathematics is higher dimensional and is similar to the mathematics of the three dimensions of space.

Six-dimensional space is reality, the world as it is. This reality we perceive only through the slit of our senses, touch and vision, and define as three-dimensional space, ascribing to it Euclidean properties. Every six-dimensional body becomes for us a three-dimensional body *existing in time*, and the properties of the fifth and sixth dimensions remain for us imperceptible.

P. D. Ouspensky, A New Model of the Universe (page 426)

16

The Nature of Reality

Scientists say that the universe we live in is composed of atoms, which are further composed of protons, neutrons, and electrons. These are further composed of tinier subatomic particles like quarks and others.

On the smallest level, the subdivision of subatomic particles continues till we reach the realm of super strings. These are tiny vibrating strings that give rise to subatomic particles and all matter.

However, deep down it all comes down to information. The ultimate constituent of our universe and the true nature of reality is information. Everything you see and touch is composed of information. Everything that enters our brain through the five senses is converted to electrical signals in the brain which are basically information.

Some scientists believe that we live in a simulation, i.e. a simulated reality or a programmed universe like a movie or a video game. In particular, some believe that we live in a MMORPG (Massively Multiplayer Online Role-Playing Game).

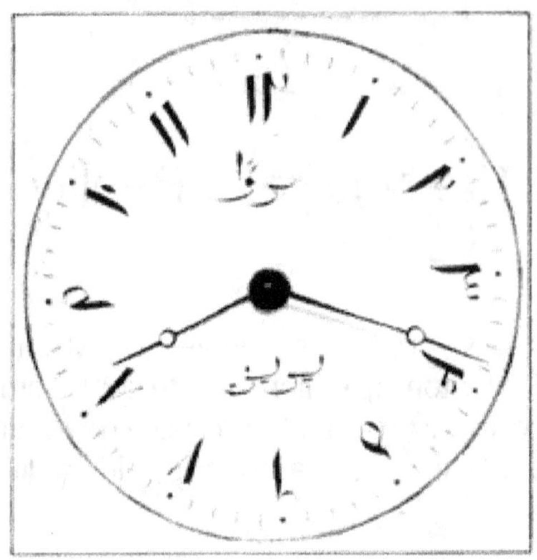

… every living creature is equipped at birth with a … clock, which is set by the remorseless hand of racial ancestry for a limited number of days or months or years. The changes and chances of this mortal life may stop it before its time, and exceptionally favorable conditions may slightly prolong its beat; but in normal circumstances that internal clock will run for a precise period determined by heredity and no longer.

A. Wyatt Tilby, The Quest of Reality (page 177)

17

Consciousness

What is consciousness? This is what we are. It is our awareness of who we are and what we feel and do.

During the past few years scientists have been studying consciousness extensively in the hope of arriving at a physical (i.e. mathematical) theory of consciousness.

Consciousness entered the world of science and physics through the subject of quantum mechanics. Scientists found out that the consciousness of the observer plays a crucial role in determining the outcomes of experiments and decisions. Some scientists further conclude that we create our own reality.

It is not clear how consciousness originates. Some say it emerges due to the complexity of the human brain. Others maintain that consciousness is the primary reality and everything else including matter comes out as a result of consciousness.

It has been mentioned previously in this book that time is a construct of consciousness.

If one event is linked with another in time, then we are dealing with memory: the former event is in some sense "remembered" by the latter.

Ervin Laszlo, The Whispering Pond (pages 172)

18

Healing Yourself

There is a method in which you can heal yourself or rejuvenate yourself using techniques that are based on time and time manipulation.

Using such techniques, to heal yourself you need to become your younger self again. This can be done by doing mental time travel to the past – to a time that you were particularly vibrant with energy. Your current self can then use the extra energy of your younger self to rejuvenate yourself in the present. This technique also has the added benefits of healing all types of diseases. It can even restore lost youth and vibrant energy.

This technique coupled with love would constitute the fountain of youth – the elixir of life. This is what people have sought for thousands of years – it is in your capacity if you can master the technique of mental time travel to the past.

The healing effect may spill over to others and you will find out that you can heal others as well.

... the self has fragmented itself from the general field of consciousness and has become blocked from creativity so that a synchronicity now appears to be a rare and isolated incident, rather than one aspect of the general order of time and unfoldment.

F. David Peat, *Synchronicity: The Bridge Between Matter and Mind*
(page 234)

19

Synchronicity

When you see a lion on the road and then you think of a lion, this is a normal sequence of events. The cause (seeing the lion) occurs before the effect (thinking of the lion). This is the normal flow of time. Thus time goes forwards.

However, when you think of a lion first then followed by seeing a lion on the road, then this is a coincidence. In science this is called a meaningful coincidence or a synchronicity. This is not the normal flow of time. In this case, the effect (seeing the lion) occurs before the cause (thinking of the lion). Thus it appears that time goes backwards when such a coincidence occurs.

Coincidences like the above (called synchronicities) are closely linked with time. When such occurrences happen, you should know that the normal flow of time is disrupted. They also mean that you are on the correct path.

Ever since H. G. Well's 1895 publication of *The Time Machine* – which described a four-dimensional spacetime with duration a dimension like height, width, and thickness – people have been wondering why we can't travel in time as we do in space.

Clifford A. Pickover, Dreaming the Future (page 32)

20

The End of Time

Although time is a dimension, it is not like space currently. In fact, time is an incomplete space dimension. In a normal space dimension, man can travel in both directions of that space dimension. However, in the time dimension, man cannot travel in time in both directions (past and future). Man currently has no control over the time dimension. This is what we mean when we say that time is an incomplete space dimension.

Although man may never be able to understand time completely, man will eventually develop the ability to conquer time. At that time, he/she will be able to travel through time (past, present, future) like they travel in space. Time will become like space. At that point, it will be the end of time.

… Thomas Aquinas believed God to be outside of space-time and thus capable of seeing all of the universe's objects, past and future, in one blinding instant. … An observer existing outside of time , in a region called "hypertime," can see the past and future all at once.

Clifford A. Pickover, Dreaming the Future (pages 243-244)

21

The Transformation

It is believed that death is closely associated with time. The end of a person's life means the end of time for his/her personal universe or personal reality.

On a global level, the end of time will signify the end of death for all. This momentous event will be accompanied by:

End of disease

End of accidents

End of suffering and pain

End of death

The above will be accompanied with love, compassion, and kindness for all. They say that true love can bridge the gulf of time.

The idea that time can run backwards may seem amazing, but it is far from new. When you stop to think about it, any belief that time is cyclic must involve "going into reverse" at some stage, so that he world can be returned to its initial state.

Paul Davies, About Time: Einstein's Unfinished Revolution (page 219)

22

Time in the Bible

To every [thing there is] a season, and a time to every purpose under the heaven - Ecclesiastes 3:1-8.

But, beloved, be not ignorant of this one thing, that one day [is] with the Lord as a thousand years, and a thousand years as one day - 2 Peter 3:8.

So teach [us] to number our days, that we may apply [our] hearts unto wisdom - Psalms 90:12.

And sware by him that liveth for ever and ever, who created heaven, and the things that therein are, and the earth, and the things that therein are, and the sea, and the things which are therein, that there should be time no longer - Revelation 10:6.

shutterstock · 193922900

Today the subject of time travel has jumped from the pages of science fiction to the pages of physics journals as physicists explore whether it might be allowed by physical laws and even if it holds the key to how the universe began.

J. Richard Gott, Time Travel in Einstein's Universe
(page 5)

Name: ___________________________________

Age: ___________________________________

Notes

What is Time?

Facts, Musings, and Speculations About Time for Children

Time is closely linked with aging and death. In order to conquer time we must use the most powerful force in the universe: love. True love can heal the sick, cure all disease, break any spell, delay aging, restore youth, and even reverse death itself. Such a power is needed so that we may be able to control time.

What is Time?

Facts, Musings, and Speculations About Time for Children

What is Time?

Facts, Musings, and Speculations About Time for Children